DATE DUE

APR 2 9 2019	
	PRINTED IN U.S.A.

LIGHTNING
BOLT
BOOKS™

The Supersmart Orangutan

Mari Schuh

Lerner Publications ◆ Minneapolis

To St. John Vianney School

Lerner Publications Company
A division of Lerner Publishing Group, Inc.
241 First Avenue North
Minneapolis, MN 55401 USA

For reading levels and more information, look up this title at www.lernerbooks.com.

Library of Congress Cataloging-in-Publication Data

Names: Schuh, Mari C., 1975- author.
Title: The supersmart orangutan / Mari Schuh.
Description: Minneapolis : Lerner Publications, [2018] | Series: Lightning bolt books.
    Supersmart animals | Audience: Ages 6-9. | Audience: K to grade 3. | Includes bibliographical
    references and index.
Identifiers: LCCN 2017054019 (print) | LCCN 2017038554 (ebook) | ISBN 9781541525337
    (eb pdf) | ISBN 9781541519831 (lb : alk. paper) | ISBN 9781541527645 (pb : alk. paper)
Subjects: LCSH: Orangutans—Juvenile literature. | Animal intelligence—Juvenile literature.
Classification: LCC QL737.P94 (print) | LCC QL737.P94 S386 2018 (ebook) | DDC 599.88/3—dc23

LC record available at https://lccn.loc.gov/2017054019

Manufactured in the United States of America
1-44319-34565-11/1/2017

Table of Contents

Meet the Orangutan

An orangutan peeks out from rain forest trees. It is looking for food. Orangutans know which trees have ripe fruit. They remember how to find these trees.

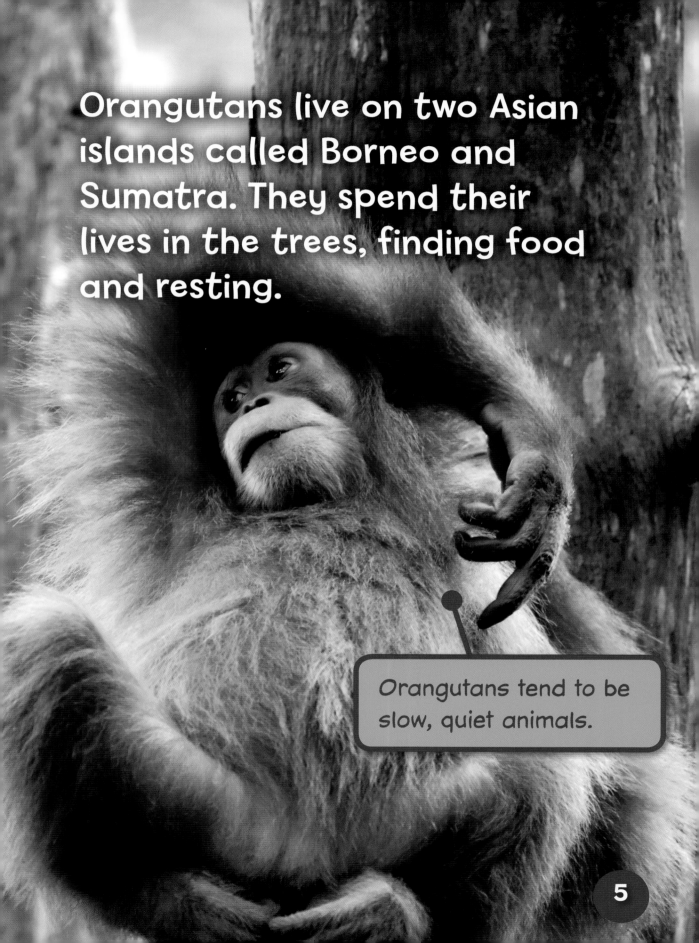

Orangutans live on two Asian islands called Borneo and Sumatra. They spend their lives in the trees, finding food and resting.

Orangutans tend to be slow, quiet animals.

Smart Orangutans

Orangutans use sticks as tools in the rain forest. Sticks help them knock fruit from trees and get honey out of beehives.

Orangutans hold things in their strong hands and mouths.

This orangutan is using a leaf umbrella!

Orangutans use leaves as gloves to hold spiny fruit. They huddle under leaves to keep out rain, just as humans use umbrellas.

Orangutans build nests in trees before sunset each night. An orangutan's nest is usually near the last place it found food. Orangutans pick strong branches to build nests.

A baby orangutan shares a nest with its mother.

Orangutans use leafy branches as blankets. Sometimes they add branches above their nest to make a roof. Some nests even have bunk beds.

This orangutan nest has just one leafy perch, but some have two—also known as bunk beds!

Sounds help orangutans communicate. Soft sounds tell babies to stay close to their mothers. Long calls can help males find females.

Scientists believe orangutans can make at least thirty-two different sounds!

Orangutans are quick learners. Some live around people and learn to do things people do. They can learn to saw wood and even paint pictures.

Scientists called primatologists study orangutans.

11

The Life of an Orangutan

Orangutans are the largest mammals that live in trees. Females weigh about 100 pounds (45 kg). Males weigh about 200 pounds (91 kg).

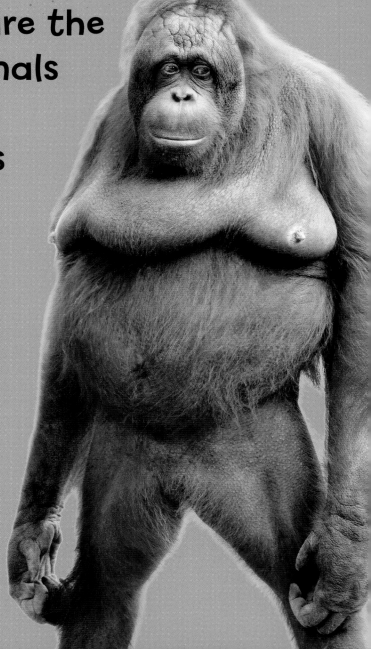

Baby orangutans cling to their mothers.

Female orangutans give birth to one baby at a time. They feed their babies fruit soon after they are born. Young orangutans drink their mother's milk for many years as they grow.

Young orangutans learn how to climb, build nests, and find food.

Young orangutans live with their mothers until they are seven or eight years old. Mothers teach their young how to live in the rain forest.

14

Orangutans grow up to have babies of their own. Females usually give birth to one baby when they are about fifteen. They give birth about every eight years.

Many orangutans live long lives. Orangutans in the wild may live more than forty years.

Orangutans in Danger

Orangutans are in danger of going extinct. Leopards, snakes, and tigers try to eat orangutans.

Leopards are predators, or animals that eat other animals.

People have cut down many trees on Borneo.

A much bigger danger to orangutans is people. Some people cut down rain forest trees. Then orangutans have less fruit to eat. People also capture baby orangutans to sell as pets. They kill the mother before taking the baby.

Yet other people are working hard to keep orangutans safe. They work to protect the rain forest. They keep rain forest trees from being cut down.

This sign protests cutting down trees.

Rescue groups help orangutans that have been kept as pets. They teach them how to live in the wild. When the orangutans are ready, they return to their home in the rain forest.

The hard work of those who love orangutans will help keep these animals safe for years to come.

Orangutan Diagram

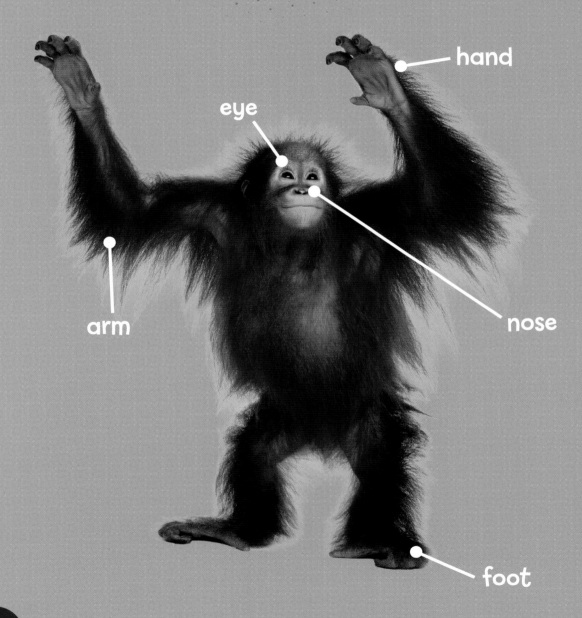

hand

eye

arm

nose

foot

Fun Facts

- Orangutans have untied boats from docks. They rode the boats across a river.

- An orangutan named Chantek learned sign language. He learned more than one hundred signs, and he even invented some of his own words!

- An orangutan named Fu Manchu picked the lock of his home at a zoo in Omaha more than once. He used a piece of bent wire, which he hid in his mouth.

Glossary

extinct: having died out

mammal: a warm-blooded animal that breathes air and has hair or fur. Mammals feed milk to their young.

rain forest: a thick area of trees where a lot of rain falls

rescue: to save from danger. Rescue groups help save captured orangutans.

ripe: fully grown and ready to be eaten

spiny: covered with many sharp, pointed parts

Further Reading

Borgert-Spaniol, Megan. *Orangutans*. Minneapolis: Bellwether Media, 2014.

Enchanted Learning: All about Orangutans
http://www.enchantedlearning.com/subjects
/apes/orangutan/

National Geographic Kids: Orangutan
http://kids.nationalgeographic.com/animals
/orangutan/

Neme, Laurel. *Orangutan Houdini: The Ape That Wouldn't Stay Put*. Piermont, NH: Bunker Hill, 2014.

Raum, Elizabeth. *Orangutans Build Tree Nests*. Mankato, MN: Amicus, 2018.

Sabatino, Michael. *Being an Orangutan*. New York: Gareth Stevens, 2014.

Index

Photo Acknowledgments

The images in this book are used with the permission of: tostphoto/iStock/Getty Images, p. 2; ullstein bild/Getty Images, p. 4; OtmarW/iStock/Getty Images, p. 5; Funny Solution Studio/ Shutterstock.com, p. 6; Jami Tarris/The Image Bank/Getty Images, p. 7; uzi Eszterhas/ Minden Pictures/Getty Images, p. 8; Debbie Monique Jolliff/Alamy Stock Photo, p. 9; GUDKOV ANDREY/Shutterstock.com, pp. 10, 13; Barcroft Media/Getty Images, p. 11; Odua Images/ Shutterstock.com, p. 12; Kjersti Joergensen/Shutterstock.com, p. 14; JohnLForeman/iStock/ Getty Images, p. 15; surassawadee/Shutterstock.com, p. 16; BAY ISMOYO/AFP/Getty Images, p. 17; Jupiterimages/The Image Bank/Getty Images, p. 18; MrMiagi/iStock/Getty Images, p. 19; Eric Isselee/Shutterstock.com, p. 20; Lillian Tveit/Shutterstock.com, p. 22.

Front cover: Sergey Uryadnikov/Shutterstock.com.

Main body text set in Billy Infant regular 28/36. Typeface provided by SparkType.